Invisible

to

Visible

Arthur L Saxton

Published by Profound Publishing LLC

Louisville, KY

ISBN: 978-0-9796146-9-9

Library of Congress Control Number: 2026903092

Cover Design

By

Arthur L. Saxton

and

Some Interior graphics and the motifs are AI Generated

Dedication

To my parents, John and Alice Saxton — whose love lit the path.

Though they had little formal education, they gave my brother and me the greatest gift of all: the Golden Rule of Life — "Do unto others as you would have them do unto you." What they lacked in schooling, they more than made up for in the wisdom and love that shaped our lives.

And the light of that path still burns.

Thanks Mama & Daddy

CONTENTS

TABLE OF CONTENTS

PREFACE

From the moment we enter this world, we are connected to a reality that is difficult to explain. Who hasn't stood in the quiet of twilight and sensed something present yet unseen? Long before I understood science or religion, I was already living in that overlap—unaware, yet shaped by it.

As a child, the unseen world fascinated me. Conversations at home often drifted toward ghosts, spirits, and mysteries that stirred my imagination. When television arrived, shows like The Invisible Man deepened my curiosity. Growing up in a small town, I spent much of my time outdoors, and the wind became my first teacher of the invisible. I could feel it, hear it, watch its effects—yet never see it. If the wind was real but unseen, what about other forces like the soul, spirits, or God? If the wind revealed itself through its effects, might deeper realities do the same?

Other childhood moments added to this quiet awakening. I remember lying on a dirt road watching ants, struck by how small and vulnerable they were. Later, as I learned about the vastness of the universe and the fragility of human life, I wondered whether we too are subject to forces far beyond our control. One of my earliest memories is watching dust dance in a beam of sunlight from my crib.

Years later, I imagined those particles as stars and planets. Even then, I sensed that great truths are often hidden in plain sight.

These early experiences planted the questions that shape this book. My hope is to show that the layers of meaning around us are available to anyone who is awake, curious, and willing to see the extraordinary woven into the ordinary. I invite you to join me—to consider the invisible as real and essential to our understanding of existence, and to search with me for signs of the Divine in both the seen and unseen.

With these memories as our compass, let us begin the journey from the invisible to the visible.

INTRODUCTION

Science asks how. Faith asks why. This book asks both, trusting that the search itself reveals something deeper. As they explore the complex world we live in, science and faith often seem to move along separate paths, yet they eventually converge, each illuminating what the other might miss.

This book is divided into three sections. The first looks at creation as told in the Bible. The second examines scientific discoveries about our origins. The third brings these perspectives together, showing how they often reflect one another at a deeper level.

The wonder of creation—with its grandeur, complexity, and mystery—demands both awe and thought. Though the Bible and science use different methods, both are captivated by the unseen beginnings of everything. Our understanding of the universe, matter, and spirit continues to grow, revealing new layers of meaning.

Yet profound questions remain. Why are humans and animals paired? Why is procreation so mysterious? What are the chances that two sexually compatible beings could appear by chance? When powerful computers calculate the odds, the staggering improbability— 10 to the 40,000th power—challenges the idea of mere randomness. This alone suggests that existence, with its complexity and purpose, points to an underlying order that arises from the invisible.

When I speak of the invisible, I mean not only the very small—atoms, electrons, quarks—but also the vast and immeasurable: the fabric of space, the nature of consciousness, and the presence of God. The unseen includes immaterial energies, spiritual realities, and forces beyond our direct perception.

This book uses "invisible" to describe things that, though hidden by size or nature, are undeniably real. How do we perceive God, who is everywhere, when His very presence makes Him indistinct within the world He fills? How do we begin to understand what lies beyond our universe, or even within the 95% of it that remains unknown to science? As finite beings navigating infinite realities, some truths will always be beyond our senses, yet clear to reason and faith.

Recognizing the unity of purpose between science and Scripture, I invite you to look past superficial differences and explore the questions that truly matter. May this book broaden your horizons and encourage you to pursue mysteries that seem beyond understanding. Together, let us seek the invisible, the eternal, and the very essence of our being.

PART ONE

BIBLICAL VIEW OF CREATION

'And the Spirit of God was hovering over the waters.'

(Genesis 1:2)

Part One The Biblical View of Creation

THE BREATH SPIRAL

Creation begins not with form, but with presence — the Spirit hovering over the waters.

📖 Chapter One: The Spirit Before Time

Before creation took shape, before light pierced the void, there was Spirit. The biblical story begins not with matter, but with presence — an invisible, living force hovering over the waters, preparing the canvas of creation. In this moment before moments, God's Spirit is already active, already moving, already speaking life into what does not yet exist.

INVISIBLE MADE VISIBLE — IN GOD'S IMAGE

Creation unfolds through intention, not accident. Genesis tells us that "the Spirit of God was hovering over the waters." Before form, color, or structure, God's invisible presence is already shaping reality. When God speaks, "Let there be light," His word is not sound alone — it is power. The Spirit energizes the command, and light appears. Order rises from chaos. Life begins.

At the height of creation, God's invisible work becomes deeply personal. God forms Adam from the dust — visible matter — but Adam does not live until God breathes into him "the breath of life." That breath is Spirit: unseen, yet essential. Humanity becomes a union of the visible and the invisible.

Scripture affirms this mystery again and again. Colossians 1:15–16 declares Christ as "the visible expression of the invisible God," and Romans 1:20 reminds us that God's invisible attributes are clearly seen through what has been made. Science, too, will echo this truth: the deepest realities are often hidden from sight.

God's Gift of Free Will

Free will is another invisible force woven into creation. In Eden, Adam is given choices — a boundary that makes love possible. This boundary is not merely about obedience; it is the space where free will becomes real.

Here, unseen forces begin shaping human experience: temptation, conscience, desire, and the quiet prompting of God's Spirit. Free will is not random preference. It is the sacred arena where humans respond to God or resist Him.

Every decision we make echoes this invisible drama. Beneath every choice lies a spiritual interplay — motives, influences, and the gentle voice of God's Spirit. In this unseen space, morality, love, faithfulness, and rebellion take shape.

In the beginning, the unseen moved.

All that followed was its echo.

Choice and Consequence

The story of humanity does not begin with failure, but with freedom. In Eden, God creates a world where love is possible because choice is possible. The garden is not merely a place of beauty — it is a space where invisible forces shape visible outcomes. Desire, trust, obedience, deception, and spiritual influence all move beneath the surface long before any action is taken.

📖 Chapter Two: Free Will and the Garden Drama

Strife, Spirit, and the Open Wellspring of Hope

The story of the fall is not merely a moment of disobedience — it is the unveiling of an invisible conflict already stirring beneath the surface of Eden. Adam and Eve stand at the crossroads of desire and trust, their hearts pulled between the voice of God and the whisper of the serpent. The temptation is subtle, not a shout but a suggestion, reshaping their perception before it reshapes their actions.

When they reach for the forbidden fruit, the rupture is immediate. Innocence collapses. The harmony of creation trembles. Sin enters quietly, yet its presence reverberates through every corner of the human soul. Fear awakens where there was once only fellowship. Shame settles where there was once only openness. The first couple, who once walked freely with God, now hide from the very presence that gave them life.

Yet even here — in the shadow of failure — the Spirit does not withdraw.

He seeks.

He calls.

He covers.

God's pursuit of Adam and Eve reveals a truth that will echo through all of Scripture: judgment may expose, but mercy restores. The garments God provides are more than coverings; they are symbols of grace, a visible sign of an invisible promise that redemption will one day come through the offspring of the woman.

This promise is spoken in the first prophecy of Scripture:

> *"I will put enmity between you and the woman,*
>
> *and between your offspring and hers;*
>
> *he will crush your head,*
>
> *and you will strike his heel."*
>
> *(Genesis 3:15)*

Here the great tension of the ages is declared. Evil will wound, but it will not win. The serpent will strike, but the Redeemer will crush. The invisible battle that began in Eden will unfold across generations, yet the outcome is already foretold.

This same tension surfaces again in the story of Cain and Abel. God warns Cain with striking imagery: "Sin is crouching at your door; it desires to have you, but you must master it." Sin is portrayed not as an abstract idea but as a living force — lurking, waiting, pressing toward destruction. The Spirit, meanwhile, urges Cain toward self-control, humility, and righteousness. But Cain yields to jealousy, and the first murder stains the ground.

Even so, the Spirit does not abandon humanity.

He continues striving, calling, and guiding.

He continues planting hope in the soil of brokenness.

Every human life becomes a stage where this ancient drama plays again — the whisper of temptation, the pull of desire, the quiet pursuit of God. Where sin seeks to enslave, the Spirit seeks to restore. Where shame bends the soul inward, grace lifts it toward healing. Even in our failures, the promise spoken in Eden remains alive, moving through history like a hidden stream beneath the surface of human experience.

Though Adam and Eve are barred from the Tree of Life, they are not barred from God's mercy. The Spirit continues His work through invisible means, preserving hope for every generation. No matter how fractured a life becomes, the wellspring of redemption remains open, waiting for the heart that turns toward it.

THE MARCH OF REDEMPTION – GOD'S SPIRIT GUIDING HISTORY

From the moment humanity steps outside God's design, the invisible Spirit of God begins a long and patient march through history — guiding, restraining, whispering, and shaping events in ways the world cannot see. Redemption does not begin at the cross; it begins in the quiet, unseen movements of God's Spirit long before humanity understands its need.

Consider Noah. While violence and corruption spread like a shadow across the earth, the Spirit stirs one man's conscience. Noah listens to a voice no one else hears. The Spirit guides his hands to build an ark — a task that seems absurd to the watching world. Yet Noah obeys, and through that obedience, the future of humanity is preserved. When the waters rise, it is not the ark alone that saves him, but the unseen hand of God carrying him through judgment into a new beginning.

Then comes Abraham, called to leave everything familiar and walk toward a promise he cannot yet see. There are no signs in the sky, no maps, no assurances from the world around him — only the persistent, invisible voice of God urging him forward. With that call comes a promise: that he and Sarah, barren and aged, will become the parents of a great nation. Their laughter, their doubts, their long waiting — all of it becomes the soil in which faith is formed. When Isaac is finally born, he is more than a child; he is a living testimony that God's unseen Spirit moves through impossibility.

But the Spirit's work does not end with Isaac's birth. It deepens in the moment Abraham is asked to offer his son on the mountain. The command pierces the heart, yet Abraham obeys, believing that the God who gives life can also preserve it. At the final moment, a ram appears — a visible sign of an invisible mercy. This scene foreshadows a greater mystery: that God Himself will one day provide His own Son for the redemption of all.

Generation after generation, the Spirit continues guiding His people, often unnoticed but always present. Joseph's dreams are not mere imagination — they are Spirit-given visions that rescue nations from famine. Moses encounters God in the burning bush, but it is the Spirit who strengthens him, who speaks through him, who empowers him to confront Pharaoh. Behind every visible act — the parted sea, the manna, the plagues — is an invisible process, a man carried, instructed, and sustained by God's Spirit.

The prophets carry this pattern forward. Isaiah, Ezekiel, Daniel — each sees visions too vivid to be born of human thought. Chariots of fire, dry bones rising, a suffering servant who will bear the sins of many. These visions are not merely revelations; they are Spirit-breathed windows into God's unfolding plan. Through them, the invisible Spirit continues shaping history, preparing the world for redemption.

And then, in the fullness of time, the Spirit's march reaches its turning point. A child is conceived by the Holy Spirit, born in Bethlehem, and laid in a manger. The promise whispered in Eden, carried through Noah, Abraham, Moses, and the prophets, now takes on flesh. The invisible becomes visible. The long march of redemption steps into the world through Jesus Christ.

Through all of this, one truth becomes unmistakable: salvation is never the product of human effort or chance. It is always the outworking of God's invisible presence and power. History is not a random sequence of events; it is a tapestry woven by the Spirit — sometimes quietly, sometimes dramatically, but always purposefully.

The visible world reflects the invisible will of God.

And the march of redemption continues still.

The Spirit's Presence Before All Things

The thread running through Scripture is unmistakable: long before human eyes perceive movement, the invisible Spirit of God is already at work. History does not unfold by accident or by the strength of kings. It unfolds because the

Spirit moves — quietly, persistently, and with purpose.

Pharaoh resists Moses with hardened resolve, yet even his defiance becomes a stage upon which the Spirit displays God's power. What looks like political stubbornness is, in truth, the Spirit guiding Israel toward deliverance. The plagues, the Passover, the parting of the sea — each visible act is rooted in an invisible orchestration that no earthly ruler can hinder.

Nebuchadnezzar rises to dominate nations, yet his empire ultimately bends toward a divine purpose he cannot see. His dreams, his madness, his restoration — all are shaped by the Spirit who governs kings even when they do not acknowledge Him. Babylon's ascent and fall are not random; they are woven into a larger tapestry the Spirit has been crafting since Eden.

Even Rome, the empire that crucifies Jesus, becomes an unwitting instrument of redemption. Its roads, its language, its political reach — all become channels through which the gospel spreads. The empire believes it is expanding by military might, but the Spirit is preparing a world ready to receive the visible Christ.

Human rulers wield power, but they do not direct history.

The unseen Spirit does.

He bends the arc of nations toward redemption.

He fulfills prophecy spoken centuries before.

He sustains creation with forces the world cannot measure.

He guides events long before they unfold.

Just as gravity holds planets in orbit, just as electromagnetism binds atoms together, the Spirit holds the story of humanity in place, ensuring that nothing — not rebellion, not empire, not evil — can derail the purposes of God.

Creation did not end with light or land or the forming of boundaries. It continued with the shaping of lives — ordinary people drawn into extraordinary purpose by the quiet movement of God's invisible Spirit. David's story is one of Scripture's clearest portraits of how the unseen Spirit chooses, forms, and empowers a human life from the inside out.

Grace walks ahead of every fall,

preparing the place where we rise.

📖 Chapter Three: David and the Spirit's Empowering Presence

🐑 👑 MOTIF-RISE: THE SPIRIT'S ROLE IN DAVID'S UNLIKELY RISE

David's rise begins not with human recognition but with the Spirit's hidden choosing. When Samuel arrives at Jesse's house to anoint Israel's next king, every visible sign points to David's brothers — their stature, their strength, their presence. Yet the Spirit whispers a different truth:

> "People look at the outward appearance,
>
> but the Lord looks at the heart."
>
> (1 Samuel 16:7)

This is the moment the invisible becomes decisive.

David is not chosen because he is impressive; he is chosen

because the Spirit sees what others cannot.

When Samuel anoints him, Scripture records a turning point:

> *"The Spirit of the Lord came powerfully upon David from that day forward."*
>
> *(1 Samuel 16:13)*

From that moment, David's life is carried by an unseen current.

His courage, his creativity, his leadership — all of it flows from the Spirit's quiet empowerment. The world sees a shepherd boy. Heaven sees a king in the making.

David's rise is not a story of human ambition.

It is the story of the Spirit lifting the humble, shaping a heart that remembers Eden's peace, and preparing a life that will carry God's purposes into history.

THE SPIRIT'S POWER IN OVERCOMING GOLIATH

David's confrontation with Goliath is often told as a tale of bravery, but beneath the surface lies a deeper truth: this is a clash of invisible forces.

Goliath stands in visible strength — armor gleaming,

voice thundering, confidence rooted in size and steel. David stands with nothing but a sling, a few stones, and the unseen Spirit who has been shaping him in solitude.

Before the battle, David declares:

> *"You come against me with sword, spear, and javelin,*
>
> *but I come against you in the name of the Lord*
>
> *Almighty."*
>
> *(1 Samuel 17:45)*

This is not bravado.

It is awareness — the awareness that the Spirit who guided his hands against lions and bears now guides him against a giant.

When the stone flies, it is more than skill.

It is Spirit-directed precision, the invisible hand of God turning a shepherd's courage into a nation's deliverance.

The victory is not David's alone.

It is the Spirit revealing that the unseen is stronger than the seen, and that faith aligned with the Spirit collapses realities that fear cannot face.

The Spirit's Guidance in His Kingship

David's journey to the throne is not immediate.

It is shaped through caves, deserts, betrayals, and long seasons of waiting — each one a classroom where the Spirit forms the inner life of a king.

While Saul rules with visible authority, David is ruled by the invisible presence he refuses to lose. When he sins, his cry reveals what he values most:

> *"Do not take Your Holy Spirit from me."*
>
> *(Psalm 51:11)*

David knows the truth many leaders forget:

a throne without the Spirit is only a chair.

Throughout his reign:

- the Spirit guides his decisions
- the Spirit convicts his conscience
- the Spirit inspires his psalms
- the Spirit restores him when he falls
- the Spirit anchors him when he is pursued
- the Spirit reminds him of the peace humanity once knew in Eden

David's kingship becomes a living testimony that the

invisible Spirit is the true source of wisdom, courage, and restoration.

And through David, the Spirit prepares the world for a greater King — one who will not merely be empowered by the Spirit, but conceived by Him, filled with Him, and poured out through Him.

Divine Direction

The Spirit's movement through creation did not end with the shaping of nature or the ordering of the cosmos. It extended into the shaping of nations, rulers, and empires. As history unfolds, we discover that God's sovereignty is not confined to those who acknowledge Him. His invisible hand guides the faithful and the unaware alike, weaving His purposes through the rise and fall of kingdoms.

This chapter invites us into a realm where prophecy precedes politics, where divine intention outruns human ambition, and where the unseen Spirit directs even the most unlikely vessels toward His redemptive design.

The Spirit walks ahead of every story,

even when the path is hidden.

The Spirit who shapes a single life

also shapes the rise and fall of nations.

What He begins in the quiet heart,

He completes upon the stage of history.

📖 Chapter Four: God's Sovereignty Through Unexpected Vessels

🏛 The Spirit's Hand in History: Isaiah's Prophecy of Cyrus the Great

God's sovereignty is not limited to those who worship Him. He moves through kings who do not know His name, through empires that do not honor His covenant, and through rulers who believe they act by their own strength. Nowhere is this truth more striking than in the prophecy of Cyrus the Great.

More than a century before Cyrus rose to power, Isaiah recorded his name — a future ruler who would free Israel without ever realizing he was fulfilling divine prophecy. When the moment arrives, God speaks directly to this pagan king:

> *"I am the Lord, and there is no other; apart from me there is no God. I will strengthen you, though you have not acknowledged me."*

Here, God reveals a profound truth:

His sovereignty does not depend on human recognition.

Cyrus ruled with visible authority, but he walked a path shaped by an invisible hand. His decrees aligned with a divine purpose he never perceived. His victories unfolded according to a script written long before he was born. Through Cyrus, God demonstrates that His will cannot be obstructed — not by unbelief, not by empire, not by human ignorance.

The Spirit moves where He wills, even through those who do not know Him.

B. ⊛ The Invisible Architect: God's Spirit Moving Beyond Human Awareness

When Cyrus conquered Babylon in 539 B.C., Isaiah's prophecy had already been waiting in the scrolls for generations — a divine decree poised to unfold at the appointed time. Without knowing the God of Israel, Cyrus issued an edict allowing the Jewish exiles to return home and rebuild the temple (Ezra 1:1–4). In doing so, he stepped into a prophecy spoken more than a century earlier, guided by a hand he could not see.

This moment reveals a truth that echoes through all of Scripture:

God does not require human awareness to accomplish His will.

His sovereignty moves through history with quiet authority, directing rulers, nations, and events toward a higher purpose. Cyrus acted with political power, yet his decisions aligned perfectly with God's preordained plan. The Creator directs the world beyond human perception, weaving His intentions through the choices of kings who believe they act alone.

Though Cyrus never knew the God who guided his steps, his life reveals a pattern repeated throughout Scripture:

- God moves through the faithful

- God moves through the unaware
- God moves through the unwilling
- God moves through the unexpected

Kingdoms rise, empires fall, and rulers make decisions for reasons they believe are their own — yet behind every moment stands the unseen Spirit, directing the flow of time toward redemption.

If God can guide a king who does not know Him,

how much more can He work through those who seek His presence?

With this truth before us, we now turn from the movements of nations to the movements of the human heart — where the unseen Spirit continues His work in ways just as profound, just as unexpected, and just as sovereign.

Incarnation Light

Creation did not end with the first burst of light. It continued with the shaping of sky and sea, the division of waters, and the forming of boundaries. In these movements we see not only structure, but sacred rhythm — the invisible Spirit carving space into purpose. Now, that same Spirit moves again, not in the vastness of the cosmos, but in the quietness of a young woman's womb. The God who formed galaxies now forms a Child.

📖 Chapter Five: The Spirit and the Birth of Christ

🕊 The Spirit's Role in the Birth of Jesus

Throughout history, God's Spirit has been the unseen force guiding redemption, but in the birth of Jesus His presence becomes profoundly intimate. When the angel Gabriel appears to Mary, he announces that her child will not come through natural means, but through the direct action of the Holy Spirit:

> *"The Holy Spirit will come upon you,*
>
> *and the power of the Most High will overshadow you."*
>
> *(Luke 1:35)*

Here the Spirit is the active agent of divine life — just as

He hovered over the waters at creation, He now hovers over Mary's womb to initiate a new beginning. The same unseen presence that shaped the cosmos now shapes the incarnation. The Creator enters creation through the breath of His own Spirit.

Joseph, too, encounters the Spirit's work. Troubled and uncertain, he resolves to quietly separate from Mary, but God intervenes through a dream:

> *"Do not be afraid to take Mary as your wife,*
>
> *because what is conceived in her is from the Holy Spirit."*
>
> *(Matthew 1:20)*

This echoes a familiar biblical pattern: God often speaks through dreams, as He did with the first Joseph in Genesis, guiding hearts through unseen wisdom. In this way, both Mary and Joseph experience the invisible prompting of the Spirit — one through miraculous conception, the other through divine reassurance.

This moment marks a new era. The Spirit who once breathed life into Adam now brings forth the Second Adam, Jesus Christ (see 1 Corinthians 15:45). The unseen reality of God becomes visible, not only through Jesus's presence, but through the way the Spirit moves within human hearts to fulfill God's plan.

The incarnation is both a miracle of flesh and a miracle of Spirit —

the invisible becoming visible,

the eternal stepping into time,

the Creator entering His creation through the quiet work of the Holy Spirit.

✨ THE CULMINATION: CHRIST, THE UNVEILING OF THE INVISIBLE IN FLESH

The story reaches its peak in the New Testament. Jesus is announced as the ultimate expression of the invisible becoming visible. His conception occurs through the action of the Holy Spirit — the divine breath once again bringing forth life.

Throughout His ministry, Jesus's actions, teachings, and miracles are described as being empowered by the Spirit of God. When He reads from Isaiah in His hometown synagogue, He declares:

> *"The Spirit of the Lord is upon me."*
>
> *(Luke 4:18)*

In this moment, Jesus boldly reveals His identity and mission as the visible embodiment of God's invisible power and purpose.

Every moment of Jesus's life is filled with meaning. His death and resurrection align with ancient prophecies and Jewish festivals, showing that what happens visibly is part of a larger, invisible plan. The resurrection is the clearest example: the Spirit not only restores life to Jesus's body but opens a new era in which death is defeated.

Here, the pattern of the invisible making itself visible reaches its triumph.

This has profound implications for anyone who follows Jesus. His life is not merely a story to imitate, but an invitation to experience the same invisible Spirit that shaped creation and guided history. By trusting in Christ, believers receive a new relationship with the Spirit and become participants in the divine plan.

📖 Chapter Six: Pentecost and the Spirit Within

The Spirit who once moved upon prophets and kings now moves within every believer. Pentecost marks a new chapter — not merely a moment of divine visitation, but the dawn of divine indwelling. Here begins the age of the Spirit within, where the invisible power of God fills human hearts and transforms ordinary lives into vessels of eternal purpose..

God With Us, The Spirit Now Indwelling Flesh

After Jesus ascends into heaven, the presence and work of God's Spirit take on a dramatic new form. No longer is the Spirit reserved for select individuals, isolated moments, or extraordinary callings. At Pentecost, the Spirit is poured out upon all believers, filling them from within. This marks a turning point in redemption's story: the invisible power of God now dwells inside those who trust Him, enabling them to live boldly, love deeply, and rise above barriers of culture, language, and fear.

The early church becomes a living, visible expression of God's Spirit on earth. Miracles, acts of compassion, unexpected courage — these are not merely natural abilities or personality traits, but evidence of the Spirit's presence. Scripture teaches that the fruits of the Spirit — love, joy, peace, patience, and more — are not produced by human effort alone, but by God's unseen work within the believer.

Even today, this remains true. When forgiveness is extended to enemies, when hope endures through suffering, when generosity flows beyond what seems reasonable, these actions testify to a greater Spirit at work. The community of faith becomes a visible sign of the invisible harmony God intended from the beginning — a reflection of creation's order now restored within human hearts.

The same Spirit who filled the early church continues His quiet work in us today, shaping hearts just as He once shaped creation. Every act of courage, every moment of forgiveness, every spark of hope bears witness to His presence within. Yet the Spirit's work does not end with empowerment — it points us toward a greater horizon.

The One who moved over the waters,

who overshadowed Mary,

who filled the disciples at Pentecost,

now prepares all creation for its final renewal.

With this truth before us, we lift our eyes toward the ultimate victory — the day when the Spirit who dwells within us will complete the restoration God promised from the beginning.

"The unseen breath of God is the force that turns dust into life, chaos into creation, and brokenness into redemption."

Redemptive
Bloom

The same Spirit who raised Jesus from the dead now lives in us. This truth is not merely comfort — it is commission. It calls us to lift our eyes toward the final restoration, where Eden is renewed, every wound is healed, and every promise finds its fulfillment in the presence of God.

📖 Chapter Seven: Victory and Restoration

🌿 Victory and the Restoration of All Things, A Spirit-Breathed Hope

The story of Scripture does not end in struggle or tragedy, but in hope and victory. The New Testament boldly proclaims that "the same Spirit who raised Jesus from the dead lives in you." This is not poetic encouragement — it is a declaration that the invisible power present at creation and resurrection is now active within the lives of believers.

The book of Revelation unveils the final chapter of God's plan: Eden restored, the Tree of Life offered for healing, and God's presence — His Spirit — fully revealed. What was once hidden becomes visible. What was once promised becomes reality. The invisible does not fade away; it becomes gloriously, eternally seen.

Every moment of history — the struggles, the hopes, the failures, the victories — moves toward this unveiling. The

Spirit's work is relentless and unstoppable. Every promise is kept. Every wound is healed. Every longing finds its fulfillment in the presence of God.

This hope reshapes how we live. It means our present suffering is never wasted, and our pursuit of goodness is never powerless. We live between the times — anchored in the unseen, yet looking toward the day when all will be revealed. The Spirit within us is both the guarantee of what is coming and the strength we need for the journey now.

To truly live is to see beyond the surface. The invisible is not absence — it is essence. Let us now answer the call to perceive the Spirit's movement, to live with awakened eyes, and to walk in the reality of the invisible made visible.

📖 Chapter Eight: A Call to See the Invisible

👁 A COMPELLING CALL TO SEE THE INVISIBLE AND ENTER ITS REALITY

The great challenge of our age is not the lack of information, but the loss of perception. We often limit ourselves to what can be measured, weighed, or proven, yet the world is filled with invisible realities — emotions, thoughts, consciousness, spiritual longings. To be fully human is to recognize, respect, and seek these unseen depths.

Scripture calls us to live with awakened eyes, to discern the movement of God's Spirit at work around us, and to participate in His ongoing creation. The invisible is not a realm separate from daily life; it is the foundation beneath it.

🕮 The Invisible Word Becoming Visible Scripture

From the beginning, God's voice was heard but not seen — a whisper to Abraham, a fire to Moses, a vision to Isaiah. These encounters were invisible experiences, carried by the Spirit into human hearts. Yet over time, the people of God recognized the need to preserve what had been spoken. The invisible Word had to become visible text.

This is the story of the canon.

The canon was not invented by councils or imposed by rulers. It was recognized by the worshiping community, guided by the Spirit, as the writings that carried divine breath. The Law, the Prophets, and the Writings became the visible witness to God's invisible presence. Later, the apostles' letters and gospels were received in the same way — not because they were clever or popular, but because they bore the unmistakable imprint of God's Spirit.

Other writings circulated — some wise, some imaginative, some spiritually insightful. Books like Enoch shaped the imagination of Israel and echoed through the world Jesus entered. But only those writings that carried the Spirit's authority became Scripture. Only those recognized across the church became the visible vessel of the invisible God.

In this way, the canon itself is an act of incarnation.

The unseen Word becomes visible text.

The eternal becomes readable.

The Spirit's breath becomes a lamp for every generation.

The canon is not a cage for God's voice; it is a testimony that the invisible has been made visible — preserved so that the same Spirit who hovered over the waters can still speak through the written Word today.

Studying the world scientifically does not compete with faith — it complements it. For the believer, exploring the wonders of creation can become an act of worship, a way of drawing near to the Creator. Faith and reason, Scripture and nature, are not rivals but companions, united by the same Spirit who once "hovered over the waters" and now hovers over every life.

As you continue your exploration — through spiritual practice, scientific study, or both — be convinced of this: the deepest truths are not discovered by accident, but by following the movements of God's Spirit. The invisible will be made visible. The promises will become reality.

As we move from the biblical arc of redemption into the breath that bridges Spirit and science, let us pause. The Spirit who hovered over the waters now whispers through equations and energy. The next chapter invites us to see the invisible not only in Scripture, but in the very structure of creation itself.

To see the invisible is to live with awakened eyes.

Where Spirit and Science meet, breath becomes bond, and the invisible finds its equation.

📖 Chapter Nine: The Breath Beneath the Equation

Before we shift into the scientific view of creation, we pause to reflect on the shared language between Scripture and science. Both speak of beginnings. Both describe unseen forces shaping reality. Both reveal mysteries unfolding through time. The Spirit hovered over the waters; energy surged through the void. One speaks in poetry, the other in equations — yet both point to a truth deeper than sight.

⚛ Where Spirit and Science Meet

In the biblical account, the invisible Spirit animates dust, guides history, and breathes life into the womb of Mary. In the scientific account, invisible forces — gravity, electromagnetism, quantum fields — shape galaxies, atoms, and the emergence of life. These are not opposing stories. They are parallel witnesses.

The breath of God and the breath of the universe may

seem different, yet both are unseen, both are generative, and both are essential. The Spirit moves through prophets and particles alike. The same mystery that stirred Abraham's heart also stirs the curiosity of physicists. The same invisible hand that guided David's stone also guides the trajectory of electrons.

Science does not threaten faith — it echoes it. Every discovery about the unseen structure of reality becomes another way to marvel at the Architect behind it. The more we uncover, the more we realize how much remains hidden. And in that hiddenness, we find awe.

Part One has traced the movement of the unseen Spirit from creation's first breath to the final restoration promised in Scripture. We have followed His presence through prophets, kings, nations, and the birth of Christ. We have watched Him fill the early church and dwell within believers today.

And now, standing at the threshold between revelation and inquiry, we prepare to see the invisible from another angle — not abandoning the Spirit, but following Him into the very structure of creation.

The next chapter invites us to look through the lens of science, not to replace the Spirit, but to recognize His fingerprints in the fabric of reality itself.

Part Two The Scientific View of Creation

THE QUANTUM THREAD

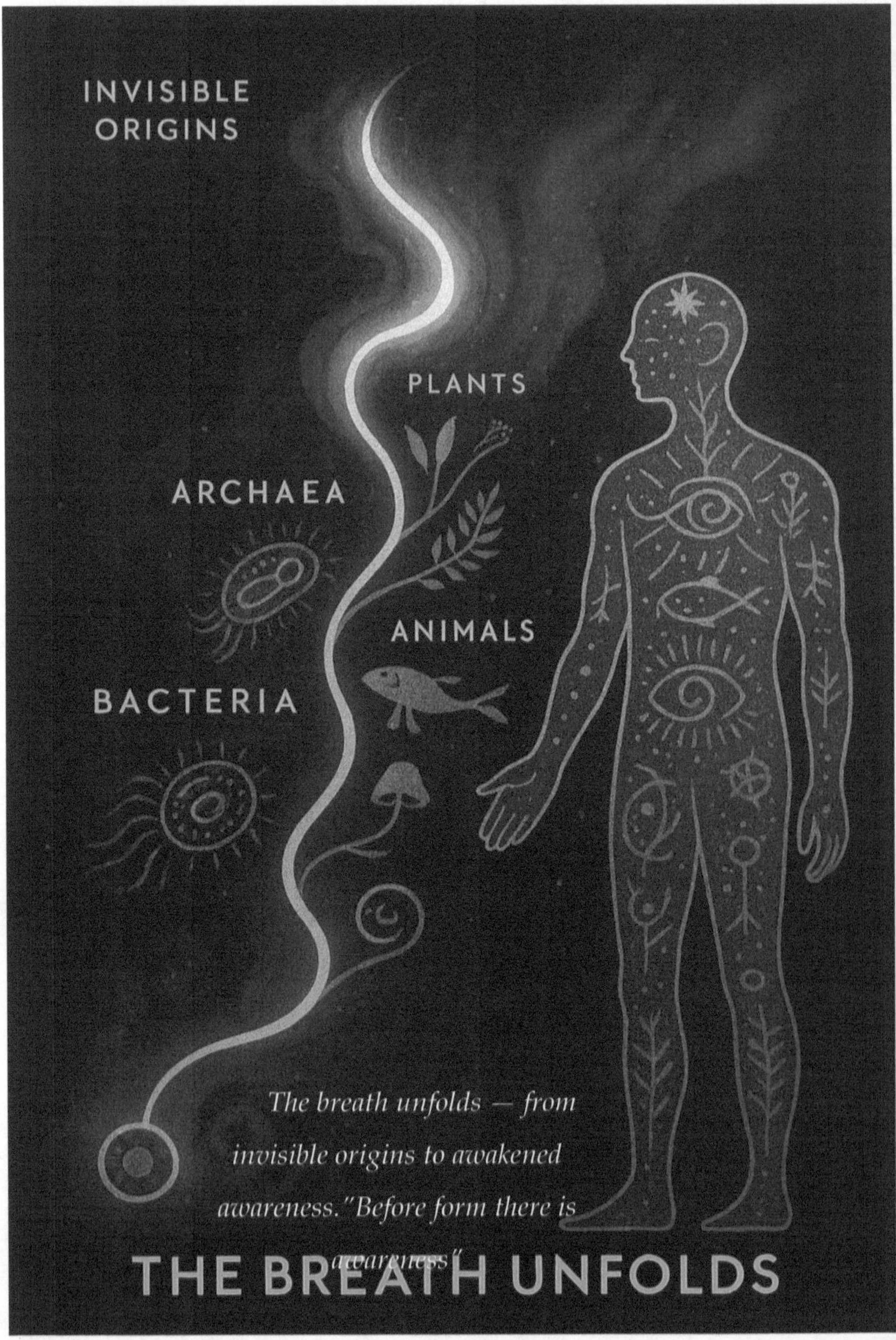
INVISIBLE
ORIGINS
PLANTS
ARCHAEA
ANIMALS
BACTERIA
The breath unfolds — from invisible origins to awakened awareness."Before form there is awareness"
THE BREATH UNFOLDS

"Let us now descend from Spirit into structure, from breath into bond."

THE UNSEEN BREATH

Before form, there is breath.
Before measurement, there is mystery,
The unseen breath moves through matter,
awakening awareness.

Chapter One: The Unseen Forces That Shape the Universe

The Spirit that hovered over the waters finds a mirror in the forces that shaped the cosmos. Science, like scripture, begins with the invisible—gravity, energy, and quantum fields. These unseen realities formed galaxies, forged stars, and continue to shape the very fabric of our lives. Let us now explore how the universe itself bears witness to the invisible.

The Unseen Forces That Shape the Universe, And Us

Science is, at its core, a search for the invisible. It seeks hidden patterns, uncovers the forces behind change, and reveals structure beneath what appears to be chaos. Again and again, discovery shows that the deepest truths — the ones that shape existence itself — are not immediately visible to the eye.

The origin of the universe is the most striking example. Physicists describe a beginning so small, so dense, so utterly beyond comprehension that it defies imagination. A fraction of a second after the first moment, the universe was smaller than an atom — a seed of pure potential, invisible yet overflowing with the energy that would become everything.

The Big Bang was not merely an explosion; it was the birth of space, time, and the fundamental forces that govern reality. As the universe expanded and cooled, order emerged from chaos through gravity, electromagnetism, and the nuclear forces. These unseen principles built the cosmos: particles formed atoms, atoms formed stars, and stars forged the elements that sustain life.

Yet these forces do not only shape distant galaxies — they are at work within us every moment.

Gravity anchors us to the earth.

Unseen electromagnetic fields pulse through every cell.

Quantum interactions influence even the smallest biological processes.

The same invisible rules that created the stars continue to shape our bodies, our minds, and our very existence in ways we are only beginning to understand.

Science may never fully grasp the first moments of creation, but its discoveries point to a profound truth: the unseen is not peripheral — it is foundational. It governs the vastness of space and the intricate workings of life itself. To study the universe is to study the invisible, and in doing so, to glimpse the structure beneath all things.

B. Breath: Where The Cosmic Becomes Intimate

Breath is where the cosmic becomes intimate. It is the meeting place of physics and personhood, the bridge between the vastness of the universe and the immediacy of your own existence.

With every inhale, atoms forged in ancient stars enter your lungs.

With every exhale, your life joins the rhythm of creation.

In this sacred exchange, the invisible becomes personal, and the structure of the universe moves through you in quiet, continual grace.

As we turn to the next chapter, we follow this breath deeper still — into the hidden architecture of life, where the quantum and the biological meet in ways both wondrous and profoundly human.

"Every atom carries a whisper of the One who spoke it into being."

📖 Chapter Two: Breath and the Quantum Bond

The invisible forces that shaped galaxies now pulse through every cell of our being. But to truly grasp their intimacy, we must descend from the vastness of space into the quiet rhythm of breath. It is here, in the sacred exchange between air and body, that the cosmic becomes personal—not just in theory, but in every inhale. Let us now enter this intimate choreography, where quantum inheritance meets the gift of life.

INVISIBLE POWERS IN BODY AND BEING: THE SYMPHONY WITHIN

Just as unseen forces shaped the galaxies, they also orchestrate the human body — a vast network of hidden interactions working in harmony. From the breath that sustains us to the electrical currents that spark thought, life itself is governed by the invisible.

The body, in its silent coordination, is more than a collection of organs; it is a living symphony conducted by forces we cannot see. The pulse of the heart, the electric whispers of muscle and mind, the constant exchange of atoms and energy — every movement is guided by invisible rhythms working in perfect unity.

Yet within this grand composition, one note repeats with sacred constancy: breath. It is the quiet solo that sustains the entire score. To understand the gift of breath is to glimpse the quantum choreography behind life itself — a dance of atoms forged in stars, now entering your lungs with every inhale.

Breath is where the cosmic becomes intimate. It is the meeting place of physics and personhood, the bridge between the vastness of the universe and the immediacy of your own existence. With every rise and fall of the chest, the invisible becomes personal.

ENTRAINMENT PARAGRAPH

Entrainment is the quiet tendency of one rhythm to fall

into harmony with another. When a stronger or steadier pattern is present, the body instinctively adjusts itself to match it. This is why heart rate slows when breathing deepens, why people walking side by side often begin moving in the same cadence, and why shared attention can synchronize brainwaves. Entrainment reveals something profound about our design: we are not isolated beings sealed off from the world. We are responsive, resonant, and shaped by the rhythms around us — drawn into connection by forces we cannot see yet continually feel.

Running Water Triggering Urination

The body is far more responsive to its environment than we often realize. Even something as simple as the sound of running water can trigger the urge to urinate. This is not coincidence; it is entrainment.

The nervous system relaxes in response to the sound of flowing water, and that relaxation lowers the body's internal resistance, allowing a hidden reflex to surface. A cue in the environment awakens a response within us. This small, almost humorous example reveals a profound truth: we are not sealed off from the world around us. The body listens, responds, and harmonizes with creation in ways that are subtle yet deeply woven into our design.

B. The Harmony of Atoms and the Gift of Breath

Each breath you take is more than the intake of air; it is an interaction with unseen forces that sustain life. The molecules you breathe — oxygen, nitrogen, argon — are bound by quantum forces forged in distant stars. Every inhale draws ancient stardust into your lungs; every exhale returns your life to the rhythm of creation.

Yet even beyond chemistry, your body responds invisibly. Sensory receptors in your brainstem detect rising carbon dioxide levels, triggering electrical signals that expand your chest before you even realize you're breathing. You do not see the intricate system governing this act, but it is precise, unfailing, and essential to life.

Breathing is the most constant reminder that the universe is not distant. It is within you. It sustains you. It moves through you with every moment. The same invisible forces that shaped galaxies now shape the rhythm of your existence.

Breath is the quiet covenant between the cosmos and the body — a continual exchange that binds you to creation itself.

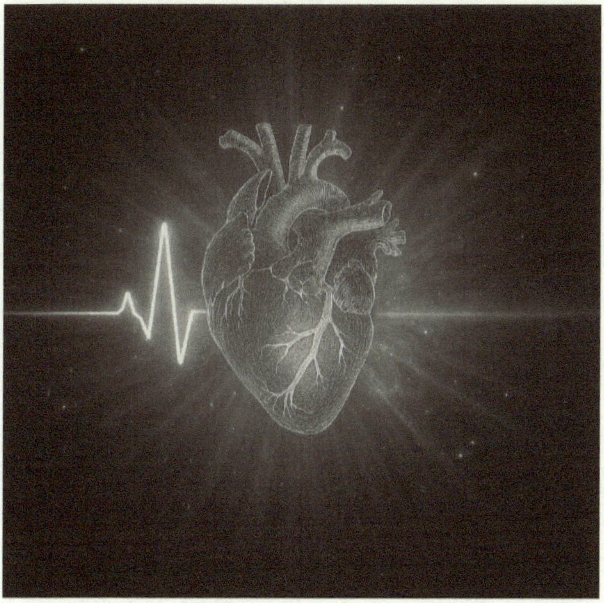

The heart's hidden rhythm:

Shaped by an unseen force

The breath that enters your lungs carries the memory of stars, but the story does not end with breath. Beneath every inhale lies a deeper mystery — the invisible architecture that makes life possible. To understand the intimacy between Spirit and science, we must descend further still, into the realm where particles flicker, probabilities dance, and existence itself is woven from the unseen.

📖 Chapter Three: The Heart's Hidden Rhythm

💓 THE PULSE OF THE HEART AND RIVERS OF ELECTROMAGNETISM

The heartbeat — the ancient drum of life — begins not with visible movement but with unseen electricity. Inside the sinoatrial node, ions quietly shift positions, triggering an electrical wave that travels through the heart muscle, coordinating every beat with perfect precision. None of this is visible, yet it keeps you alive.

Even emotions — stress, joy, sorrow, longing — alter heart rhythms without our awareness. Science can measure these changes in EKG tracings, but the true source remains hidden inside the temple of flesh. The heart listens to the invisible long before the mind understands.

The body does not wait for conscious thought to respond.

It reacts to fear before you name it.

It softens under love before you speak it.

It trembles under awe before you explain it.

The heart is the interpreter of the unseen.

THE HIDDEN CONVERSATION BETWEEN HEART AND BRAIN

For centuries, the heart was seen as the seat of emotion. Today, science reveals something astonishing: the heart and brain are in constant dialogue, exchanging signals through electrical impulses, hormones, and electromagnetic fields.

• The heart sends more signals to the brain than the brain sends to the heart.

• These signals influence attention, memory, and emotional processing.

• The heart's electromagnetic field extends several feet beyond the body.

You are surrounded by an invisible signature — a living field shaped by every beat.

This is not metaphor. It is measurable.

And yet, it is also deeply spiritual.

When Scripture speaks of God searching the heart, guiding the heart, renewing the heart, it is describing a place where the invisible and the physical meet. The heart is both instrument and sanctuary — a biological engine and a spiritual compass.

💓 THE HEART AS A LIVING METRONOME OF THE UNSEEN

Every heartbeat is a rhythm of trust.

It contracts and releases without your command.

It keeps time even while you sleep.

It beats faithfully through joy and grief, certainty and doubt.

This rhythm is not random. It is responsive.

- When you breathe slowly, the heart slows.
- When you feel peace, the rhythm smooths.
- When you pray, meditate, or rest in silence, the heart enters coherence — a state of harmony measurable in its electrical patterns.

The invisible becomes visible in the heart's response.

The Spirit's whisper often arrives not as a thought, but as a shift in rhythm — a quieting, a softening, a settling. The heart feels what the mind has not yet named.

B. THE SPIRIT'S RESONANCE IN THE HEART'S HIDDEN RHYTHM

The heart is more than a pump.

It is a place of resonance.

The same Spirit who hovered over the waters hovers over the chambers of the heart. The same invisible presence that shaped creation shapes the inner life of every believer. The heart becomes the meeting place of biology and breath, electricity and emotion, science and Spirit.

When Scripture says:

> *"I will give you a new heart and put a new spirit within you."*
>
> *(Ezekiel 36:26)*

It is not speaking only of morality — it is speaking of transformation at the deepest level of being.

The heart is where the invisible becomes intimate.

TOWARD THE NEXT MYSTERY

From the cosmos to breath, from breath to heartbeat, we have followed the invisible into the architecture of life. But the journey continues. Beyond the heart lies another frontier – the mystery of consciousness, where thought, identity, and Spirit intertwine.

The next chapter invites us into that hidden rim, where the unseen becomes the very essence of what it means to be human.

What is a thought?

It begins as a silent electrical impulse, a neuron firing, passing chemical signals across synapses, shifting electrical voltages by millivolts.

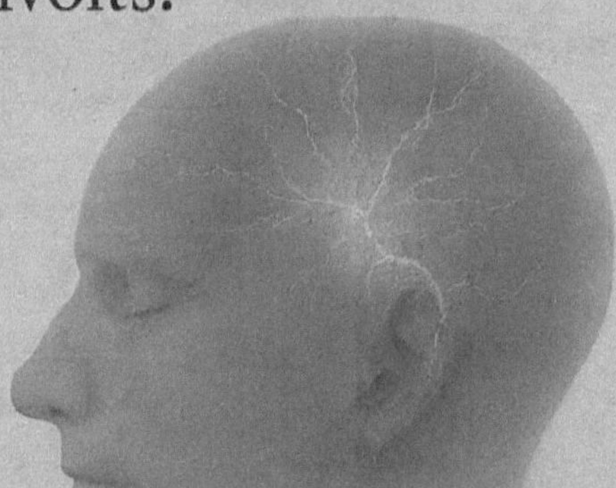

But where does memory arise? Where does the mind itself emerge.

📖 Chapter Four: The Mind Beyond Measurement

🧠✨🧠

The heart pulses with unseen electricity, but the mind dances with something even more elusive. Thoughts arise, memories form, and identity takes shape—all through signals science can trace but never fully explain. As we move from the rhythm of the heart to the mystery of the mind, we enter a realm where the invisible becomes not just functional, but deeply personal.

🧠✨🧠

🧠 Thought, Memory, and the Light That Does Not Shine

What is a thought? It begins as a silent electrical impulse — a neuron firing, passing chemical signals across synapses, shifting electrical voltages by millivolts. But where does memory arise? Where does the mind itself emerge?

Despite astonishing advances in neuroscience, science maps the echoes of thought, but not its origin. Instruments can trace the electrical ripples of a decision, the neural signatures of a memory, the patterns of attention and emotion — yet the source remains hidden.

Something unseen gives rise to personality, intuition, understanding — qualities that cannot be reduced to circuits or chemistry. We can observe the brain's activity, but not the spark that animates it. We can trace brain waves, but the soul's depth lies beyond any lens.

The mind is more than matter.

It is more than electricity.

It is more than the sum of its neurons.

It is the meeting place of the visible and the invisible — where biology becomes awareness, where chemistry becomes meaning, where the unseen breathes identity into flesh.

✨ CONSCIOUSNESS: THE INVISIBLE CENTER OF BEING

Consciousness is the great frontier — the one

phenomenon science can describe but cannot define. We know when someone is awake, aware, dreaming, remembering. We can measure brain waves, track neural activity, and map regions associated with language, emotion, and decision-making. Yet none of this explains the simple, staggering truth:

You are aware that you are aware.

No scan can reveal why a memory feels tender, why a thought carries weight, or why a moment of beauty brings tears. The machinery of the brain is visible, but the experience of being alive is not.

This is the mystery at the heart of consciousness:

- Neurons fire, but meaning arises.
- Chemicals shift, but love emerges.
- Signals travel, but identity forms.

Something more than matter is at work.

🧠 The Gap Between Brain and Self

Neuroscience can describe the brain, but it cannot locate the self. It can map the regions involved in memory, but not the meaning of memory.It can trace the circuitry of emotion, but not the reason a moment matters.

There is a gap — a sacred space — between the physical brain and the lived experience of being a person. In that space:

we choose

we believe

we hope

we pray

we love

This gap is not an absence. It is an invitation.

It is the place where the Spirit whispers, where conscience stirs, where the unseen touches the seen. The mind becomes the meeting ground between the human and the divine.

✨ The Spirit's Quiet Work Within the Mind

Scripture often speaks of renewal, transformation, and illumination — all of which occur in the mind. Not the brain alone, but the inner life where thoughts, desires, and identity take shape.

When the Spirit renews the mind, it is not merely adjusting thoughts.

It is reshaping perception.

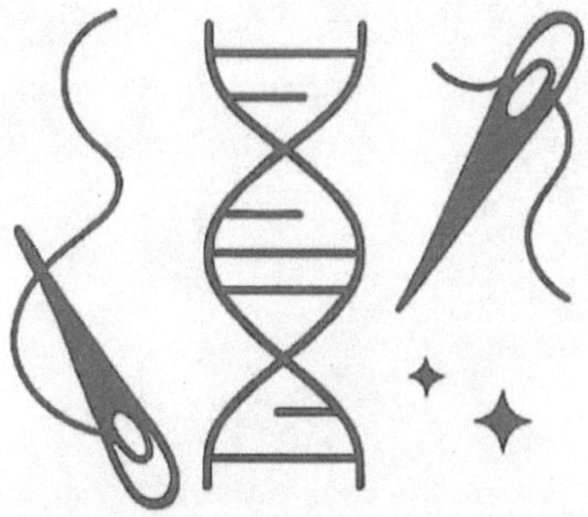

Sacred Code

Why revelation feels like remembering something we always knew.

The mind becomes a sanctuary — a place where the invisible becomes wisdom, where the eternal becomes understanding, where the presence of God becomes personal.

The mind reflects the unseen spark of consciousness, but its foundation lies deeper still — in the microscopic code of life. DNA, hidden within every cell, carries instructions written in silence. And beneath even that, quantum processes whisper through biology. Let us now descend into the blueprint itself, where the invisible becomes instruction, and life unfolds from sacred design.

📖 Chapter Five: The Blueprint of Life and Quantum Design

🧬 The Unseen Matrix: DNA, Quantum Coherence, and the Mystery of Animation

Within every cell lies a hidden blueprint — DNA, the microscopic code that carries the instructions for life itself. The double helix spirals invisibly, directing the formation of proteins, organs, and entire bodies. But DNA is more than a molecule; it is an unseen guide, orchestrating the trillion interactions that keep life moving forward.

Every strand carries a language older than speech, a script written in four letters that somehow composes the symphony of a living being. It does not shout. It does not glow. It does not announce itself. Yet from its silent instructions, life emerges with breathtaking precision.

Even here, in the quiet of the cell, the invisible rules of quantum mechanics play a role. Quantum whispers guide biology's choreography — from the efficiency of photosynthesis to the magnetic navigation of birds, from enzyme reactions to the delicate processes that allow our senses to perceive the world.

Life listens to signals we cannot see.

Life responds to forces we cannot measure with the naked eye.

Life unfolds from a realm where probability becomes pattern, and pattern becomes person.

We do not see these forces at work, but they sustain every living thing.

At every scale, the invisible gives rise to the visible —

from the stars in the sky

to the cells within our bodies.

THE HIDDEN INTELLIGENCE OF PROTEIN FOLDING

Proteins are the workers of the cell — but before they can act, they must fold. A long chain of amino acids collapses into a precise three-dimensional shape, forming the tools, structures, and messengers of life. If the folding is wrong, the protein fails. If the folding is right, life continues.

But here is the mystery:

Proteins fold with astonishing speed and accuracy – without a blueprint, without a map, without trial and error.

A chain of hundreds of amino acids somehow knows exactly how to twist, bend, and collapse into its final form. The process is so complex that even supercomputers struggle to predict it. Yet inside your cells, it happens trillions of times a day, flawlessly.

This is not randomness.

This is not chaos.

This is choreography.

A silent intelligence moves within the cell – not consciousness, but order. Not thought, but intention. A whisper of design woven into the fabric of life.

CELLULAR COMMUNICATION: THE LANGUAGE OF THE INVISIBLE

Cells do not merely exist; they communicate. They send signals, respond to threats, repair damage, and coordinate growth. They "listen" to their environment and adjust their behavior accordingly.

- Immune cells recognize invaders they have never seen.
- Stem cells decide what type of tissue to become.
- Neurons strengthen or weaken connections based on experience.
- Cells sacrifice themselves for the good of the whole.

This is not mechanical behavior.

It is relational behavior.

Life is not a machine.

Life is a conversation.

Every cell participates in a dialogue older than language — a dialogue written in chemical signals, electrical pulses, and quantum probabilities. The body is not a collection of parts; it is a community of living intelligence.

THE SPIRIT'S IMPRINT ON BIOLOGICAL ORDER

Scripture speaks of the Spirit as the One who gives life, sustains life, and breathes order into creation. When we look into the cell, we see a reflection of that truth.

The Spirit does not replace biology.

The Spirit animates it.

The invisible order within DNA, the precision of protein folding, the communication between cells, all of it echoes the same pattern:

Life is guided by an unseen presence.

Not in a way that violates science,

but in a way that gives science its foundation.

The Spirit is the quiet architect behind the coherence of life — the One who holds the blueprint, sustains the pattern, and whispers through the quantum fabric of creation.

Life is not an accident.

Life is not a coincidence.

Life is a reflection of the invisible.

THE HARMONY OF LIFE: ORDER EMERGING FROM THE UNSEEN

Life is not assembled piece by piece.

Life emerges from harmony.

Cells do not merely follow instructions; they participate

in a coordinated symphony. Proteins fold, DNA replicates, membranes communicate, and organelles cooperate — all without a conductor we can see. The order is intrinsic, woven into the very fabric of existence.

This harmony is not mechanical.

It is relational.

It is responsive.

It is alive.

From the smallest bacterium to the most complex organism, life expresses a unity that cannot be reduced to its parts. Something unseen holds the pattern together — a coherence that echoes through every level of biology.

This coherence is not imposed from outside.

It rises from within.

THE EMERGENCE OF COMPLEXITY: WHEN THE INVISIBLE BECOMES FORM

As life grows more complex, the invisible scaffolding becomes even more astonishing. Cells organize into tissues. Tissues form organs. Organs integrate into systems. Systems unite into a living being capable of thought, emotion, memory, and love.

At each stage, the pattern deepens.

At each stage, the unseen becomes more visible.

Life is not a random assembly of parts.

Life is a revelation — unfolding from the microscopic to the magnificent.

The same invisible forces that shape quantum particles shape the architecture of the body. The same coherence that guides protein folding guides the development of the embryo. The same whisper of order that animates the cell animates the whole person.

The blueprint is not static.

It is dynamic.

It is alive.

THE BRIDGE TO ELECTRIC LIFE

As we descend through DNA, proteins, and cellular intelligence, we arrive at a profound truth:

Life is animated by invisible energy.

The blueprint provides the structure,

but electricity provides the spark.

Every thought, every heartbeat, every movement begins with voltage — a current flowing through the body's hidden circuits. The same quantum rules that guide the cell now guide the electric architecture of the human being.

The transition is natural:

- From quantum coherence
- To biological order
- To electric animation

The invisible becomes instruction.

Instruction becomes form.

Form becomes life.

And so Chapter Five leads us to the threshold of a new mystery — the body as an electric temple, where spirit and science converge in living flesh.

Part Three
The Divine Weave

THE INFINITY WEAVE

Where Spirit and Science intertwine

the weave holds all things together.

THE SACRED INTERLACE

Not stitched by hands, but woven by Spirit—the sacred interlace binds heaven and earth, matter and meaning.

Every thread returns to the One

who wove the first light.

DNA guides the formation of the body, but it is electricity that animates it. Every cell pulses with voltage, every movement begins with an unseen spark. Ancient traditions spoke of subtle energies; modern science now confirms them. Let us now explore the body not just as biology, but as a temple of electric presence—where spirit and science converge in living flesh.

📖 Chapter ONE: The Body as an Electric Temple

⚡ THE NETWORK OF ENERGY: THE BODY AS AN ELECTRIC TEMPLE

Medical science confirms that the human body is more than flesh and organs, it is an electric entity, pulsing with unseen currents. Every living cell houses a voltage, every heartbeat, muscle contraction, and neural signal is activated by electric energy, unfelt, unseen, yet as real as the air we breathe.

Ancient traditions have long spoken of chi, prana, and meridian, subtle energies flowing within us. While once dismissed as mystical, modern science now explores bioelectric fields that influence healing, memory, and emotion. The pulse we can measure is only a fraction of the current flowing beneath rivers of invisible energy that orchestrate life itself.

Just as in the universe at large, we are shaped by the unseen; woven together by forces we cannot touch, yet ones that govern every moment of our existence.

⚡ The Electric Architecture of the Body

Every moment of your life is carried by electricity.

Not the kind that powers machines, but a living current woven into every cell. The body is not merely animated by electricity — it is organized by it.

Neurons communicate through voltage changes measured in millivolts.

Muscles contract because ions flow across membranes.

The heart beats because an electrical impulse travels through its chambers like a conductor guiding a symphony.

This is not metaphor.

This is biology.

Yet the precision of this electric architecture is astonishing.

Signals travel through nerves at remarkable speeds, coordinating movement, sensation, memory, and emotion. The body is a network of circuits — not mechanical, but living, responsive, and deeply intelligent.

Electricity is the body's hidden language.

It is the medium through which intention becomes action, thought becomes movement, and awareness becomes experience.

The body is not a machine.

It is an electric temple.

⚡ The Body's Electromagnetic Field: A Living Aura

Every electrical current generates a field, and the body is no exception.

The heart produces the strongest electromagnetic field of any organ, radiating several feet beyond the body. The brain generates its own field, shifting with thought, emotion, and attention.

We do not see these fields, yet they shape our experience:

- They influence how cells communicate.
- They shift with stress and calm.
- They synchronize with others during moments of connection.

- They respond to intention, focus, and emotion.

Ancient traditions spoke of an aura, a subtle energy surrounding the body.

Modern science now measures electromagnetic fields that echo those ancient descriptions.

The invisible is not imaginary.

It is simply unseen.

⚡ The Spirit's Resonance Within the Electric Temple

If the body is an electric temple, then the Spirit is the presence that fills it.

Scripture speaks of the body as a temple of the Holy Spirit — not as a metaphor, but as a reality. The Spirit does not bypass the body. The Spirit inhabits it, resonates within it, and moves through the very architecture that sustains life.

When peace settles over the mind, the brain's electrical rhythms shift.

When love awakens in the heart, the electromagnetic field expands.

When truth pierces the soul, the body responds with

clarity, alignment, and strength.

The Spirit does not replace the body's electric design.

The Spirit harmonizes with it.

The unseen presence of God moves through the unseen currents of the body, shaping perception, guiding intuition, and awakening the deepest layers of identity.

The electric temple becomes a sanctuary —

where biology meets mystery,

where energy meets presence,

where the invisible becomes communion.

⚡ The Nervous System: The Body's Sacred Pathway of Awareness

The nervous system is the body's living network of perception — a branching river of electric pathways carrying sensation, memory, intuition, and awareness. Every thought you think, every emotion you feel, every movement you make travels through this intricate web of neurons.

It is easy to imagine the nervous system as wiring.

But wiring is dead.

Neurons are alive.

They listen.

They adapt.

They remember.

They respond.

Signals travel through these pathways at remarkable speeds, leaping across synapses in bursts of electric light. The nervous system is not merely a communication network — it is the body's internal awareness, the place where the outer world becomes inner experience.

And within this living circuitry, something deeper moves.

The nervous system does not only carry information.

It carries meaning.

It carries presence.

It carries the subtle movements of the Spirit.

"When calm takes root in the mind, the nervous system softens into harmony.

When fear rises, the pathways tighten.

When love awakens, the entire system softens and opens.

The nervous system becomes a spiritual instrument —

tuned by experience,

shaped by attention,

and responsive to the quiet movements of God.

⚡ The Heart–Brain Connection: Two Electric Centers, One Inner Life

The brain is often seen as the command center, but the heart is no passive organ. It contains its own neural network — over 40,000 neurons — capable of memory, learning, and emotional interpretation. The heart sends more signals to the brain than the brain sends to the heart.

This is not poetry.

This is physiology.

When the heart is calm, the brain becomes clear.

When the heart is distressed, the brain becomes chaotic.

When the heart is aligned, the entire nervous system resonates with coherence.

Ancient wisdom spoke of the heart as the seat of understanding.

Modern science is beginning to agree.

The heart and brain form a single electric communion —

two centers, one awareness,

two rhythms, one inner life.

And within that communion, the Spirit moves.

The body, alive with electric energy, is not isolated, it is connected. Just as particles can remain entangled across space, so too, can souls remain bound by love, intention, and prayer. Let us now move from the individual to the collective, from the electric temple to the spiritual web, where unseen ties bind us across distance and time.

📖 Chapter Two: Entanglement and the Web of Spirit

🕸 Entanglement, Unity, and the Hidden Ties Between Us

One of the most astonishing discoveries in quantum physics is entanglement — the phenomenon in which two particles become so deeply linked that a change in one instantly affects the other, no matter how far apart they travel. Einstein called it "spooky action at a distance," yet today it stands as a verified scientific reality.

But what if entanglement is not merely a property of particles?

What if it is a pattern woven into the fabric of existence itself?

Across cultures and centuries, humanity has spoken of unseen bonds — prayer that reaches across miles, compassion that softens another's heart, intention that influences outcomes beyond physical contact. Once dismissed as mystical sentiment, these ideas are now being revisited with fresh curiosity. Studies on intercessory prayer, distant healing, and emotional resonance challenge the assumption that influence must be local to be real.

Consciousness behaves like a field — subtle, expansive, and interconnected.

Thoughts ripple outward.

Emotions resonate.

Love travels.

Intention reaches.

If this is true, then spirituality and science are not adversaries, but two languages describing the same hidden architecture of connection.

The thoughts you think,

the love you offer,

the prayers you whisper —

they may touch others in ways far beyond conventional understanding.

You are more deeply connected than you realize.

🕸 The Field of Consciousness: How Awareness Connects Us

If entanglement reveals that particles can remain connected across distance, consciousness suggests that minds may do the same. Awareness is not confined to the skull; it radiates, influences, and responds. Every thought carries an energy, every emotion a frequency, every intention a direction.

We often imagine consciousness as private, sealed within the boundaries of the self. But what if consciousness behaves more like a field — subtle, permeable, and shared?

Consider how you can sense another's mood before a word is spoken.

How a room can feel heavy or peaceful the moment you enter.

How compassion softens hearts, even unspoken.

How love can be felt across miles.

These are not merely poetic impressions.

They are experiences that point toward a deeper truth:

we participate in one another's inner worlds.

Modern research explores emotional contagion, neural

mirroring, and the synchronization of brain waves between people who share intention or attention. Ancient wisdom spoke of this long before science had the language to describe it.

When two people pray together, their rhythms align.

When a community grieves, hearts resonate in shared sorrow.

When love is offered, something in the other responds.

Consciousness is not isolated.

It is relational.

It is connective.

It is part of a larger web in which every life is a thread.

You are not a solitary mind moving through an empty universe.

You are a participant in a living field of awareness —

a field shaped by presence, intention, and Spirit.

🥱 Yawning Spreading From One Person to Another

Even our simplest behaviors reveal how deeply connected we are to one another. A single yawn can ripple through a room, passing from face to face without a word being spoken. Neuroscience tells us this happens through mirror-neurons — the brain's way of joining another person's state. But the deeper truth is that yawning spreads most easily among those who feel bonded: family, friends, and people who share emotional closeness. Something in us resonates with the inner life of another. Their state has become our state. This quiet, involuntary imitation is a window into a larger mystery — that human beings are not isolated minds, but participants in a shared field of awareness, where the invisible movements of one soul can stir another.

❁ The Communion of Souls: Connection Beyond Physical Presence

Human connection is not limited to proximity.

There is a depth of communion that transcends physical presence — a resonance of spirit that binds lives together even when bodies are far apart.

You have felt this before:

- the sudden thought of someone moments before they

call

- the ache of another's sorrow without being told
- the warmth of being prayed for, even unknowingly
- the sense of being held by a love that is not physically present

These experiences are not accidents of emotion.

They are signs of a deeper truth:

souls are not isolated units — they are relational beings woven into one another.

Scripture speaks of being "members of one body," a spiritual organism in which each life is a living cell. Mystics across traditions describe a shared inner life, a communion that exists beneath words and beyond distance.

When one soul rejoices, others brighten.

When one suffers, others feel the tremor.

When one prays, the web vibrates with unseen strength.

This is not metaphor alone.

It is the lived reality of a universe where connection is fundamental.

You are part of a communion far larger than yourself.

📖 Chapter Three: Healing and the Return to Resonance

🌿 🎵 🌿

Connection is not only a comfort—it is a medicine. When the body or soul falls out of harmony, healing requires restoration of invisible order. Science now sees what ancient wisdom long taught: love, forgiveness, and coherence are forces that heal. Let us now enter the sacred space of restoration, where resonance is the remedy and Spirit is the physician.

🌿 🎵 🌿

Disease, Healing, and the Return to Resonance

At its core, sickness is a disruption of order, the cell forgets its purpose, the immune system attacks itself, the mind loses its peace. The visible symptoms, pain, fever, imbalance, are only surface signals of a deeper misalignment. Ancient wisdom teaches that healing requires restoring harmony, reconnecting body, and spirit. Now, modern science is beginning to recognize this truth.

Research shows that heart rate variability, brainwave coherence, and the effects of prayer and meditation directly influence immune function. The invisible forces that sustain life must also heal it, and when harmony is disturbed, healing requires an unseen force to restore balance.

Consider love, forgiveness, and hope, each proven to have real effects on the body. Acts of kindness reduce stress hormones. Faith strengthens resilience. Positive intention accelerates recovery. These forces cannot be bottled into pills, yet they heal at the deepest level.

Could it be that the Spirit within us is the true physician? Could aligning ourselves with unseen currents, through faith, compassion, and belief, create a medicine as potent as any drug?

Healing is not just about physical restoration; it is about returning to resonance with the unseen energies that sustain us.

📖 Chapter Four: The Threshold Between Worlds

🌌🕊️🌌

Healing restores life, but even life has its boundary. Death, long feared, may be the greatest mystery of all—the moment when the invisible departs. Yet even here, science and faith converge. Let us now approach the threshold between worlds, where the soul's journey continues, and the unseen becomes eternal.

🌌🕊️🌌

🌌🕊️ Life, Death, and the Threshold Between Worlds

Even death is woven with invisibility. The precise moment life departs cannot be fixed by clock or microscope. Organs and cells persist in their functions for minutes, sometimes hours, hinting at something continuing beyond the body. Scientists observe cells clustering, reorienting, attempting repair, almost as if they follow a hidden code, a loyalty to an unseen blueprint.

Could soul be the ordering principle, the unseen energy that departs at death? Science, unable to answer, leaves the door open for spiritual truth. Ancient traditions teach that death is not an end, but a transition. Even modern research on near-death experiences reveals patterns, visions of light, a sense of expansion, a presence beyond form.

Perhaps the journey between life and death is the greatest unseen force of all, recognized in faith, hinted at by science, and never fully explained.

📖 Chapter Five: Wisdom and the Invisible Reality

🔍✨🔍

Death does not end the story—it reveals its depth.
The greatest minds have honored mystery, and the scriptures declare that the invisible is the foundation of all that is seen. Let us now ascend to the final reflection, where reason bows to wonder, and wisdom embraces the invisible as the truest reality.

🔍✨🔍

🔍 ✨ Conclusion: Reason's Highest Wisdom, Death in the Invisible Reality

The greatest minds, Einstein, Newton, Pasteur, did not seek to eliminate mystery, but to honor it. True wisdom does not reject the unseen; it respects it as foundational to existence. Science recognizes that the deepest forces in nature, gravity, electromagnetism, quantum fields, are invisible, yet they hold everything together.

So too do the forces of the spirit - prayer, conscience, intention, and love. They shape our lives as surely as unseen energies shape the universe.

You, reader, are the intersection of forces ancient and everlasting, gravity holds you to the Earth, quantum coherence illuminates your mind, and the Spirit moves within your soul. Every atom in your body was once forged in ancient stars, and every force within you was set into motion at the dawn of time.

To embrace these truths is not to escape reality, but to step into its deepest fullness.

The invisible is not the province of ignorance; it is the pathway to wisdom. It is not the last retreat of the fearful, but the first vision of the awakened mind.

Science, at its highest peak, stands united with Scripture in declaring that creation itself is the handiwork of the Invisible God, seen now, as ever, through all that has been made.

Having explored the scientific echoes of Spirit, we now enter the Divine Weave — a sacred synthesis where biblical truth and scientific insight converge. These final chapters do not choose between domains; they reveal how each thread contributes to the tapestry of meaning

What the eye cannot trace,

the Spirit has already woven.

📖 Chapter Six: Before Creation and the Spirit That Hovered

✴☾✴

Before the universe expanded, before time began, there was Spirit—hovering, waiting, breathing potential into the void. Let us now return to the beginning, where divine intention and unified forces prepared the canvas of creation.

✴☾✴

✴ Before Creation: The Unified Forces and the Spirit That Hovered

Before the universe came into being, before stars shone and atoms moved, there was a single, incomprehensible force holding everything together in perfect unity. Physics tells us that before the universe expanded, gravity, electromagnetism, and nuclear forces were all combined into one unified energy, compressed beyond measure, waiting for the moment to separate.

The Bible speaks of a similar mystery, not bound by the laws of physics, but by divine presence. Before the earth was formed, the Spirit of God moved over the waters (Genesis 1:2), not yet creating, but waiting. Since Genesis describes not only the creation of the earth but all creation, could it be that God's invisible Spirit was the external creator of what physics describes before the universe came into being? This divine presence, waiting to start the creation process, yet being omnipresent throughout creation, having complete control of every part of it through His invisible Spirit, moving over the abyss, carrying the power to set creation in motion.

✨ The Moment of Unfolding: Creation and Expansion

Then, something happened. Science describes it as the cosmic moment, the expansion, the division of forces, the great unfolding that birthed time and space. The singular energy split, gravity took hold of matter, electromagnetism lit the heavens, and nuclear forces bound atoms together, shaping the ingredients of existence itself.

In parallel, scripture speaks of the divine command, "Let there be light" (Genesis 1:3), the moment where the Spirit that hovered initiates separation, structure, and purpose. Just as the fundamental forces divided and took their role in shaping reality, so too did God speak, distinguishing light from darkness, sky from land, order from chaos.

📖 Chapter Seven: Invisible Presence, Eternal Influence

Creation was not a single moment, but a continuous unfolding. The Spirit that hovered still moves, and the forces that shaped the stars still shape our choices. Let us now explore how presence becomes influence—how the eternal touches the everyday.

🌌 🔗 Invisible Presence, Eternal Influence

Even now, these forces remain unseen, yet they govern all things. Gravity binds planets in their dance, electromagnetism pulses through every atom, nuclear interactions sustain stars, keeping their fire alive. The invisible remains fundamental.

In the same way, scripture reveals that the Spirit remains omnipresent, sustaining all life, whispering through creation, guiding history, not always seen, but always present, shaping, upholding, renewing.

Could it be that what physics describes in equations, scripture describes in revelation, the fundamental truth that all creation is born from the unseen?

Chapter Eight: Spirit and Science as Co-Laborers

The unseen is not distant—it is foundational. Beneath every law of physics lies a lawgiver, and beneath every equation, a whisper of purpose. Let us now examine the matrix of meaning, where science and scripture speak in harmony.

ONE SOURCE, MANY REALITIES

Quantum entanglement offers a striking metaphor for the way distinct realities can arise from a single, unified Source. In entanglement, two or more particles remain mysteriously connected, sharing one underlying state even when they exist in different places and express different outcomes. Their distinction does not break their unity, and their unity does not erase their distinction. In a similar way, creation itself may be understood as multiple expressions of one divine reality—visible and invisible, temporal and eternal, earthly and heavenly—each distinct, yet all originating from the same Source who holds them together. The promise of Scripture points toward this very pattern: the old creation and the new creation, the mortal body and the resurrected body, the present age and the age to come. They are not separate universes competing with one another, but different expressions of one coherent reality sustained by God. Entanglement does not explain the Trinity or the new creation, but it gives us a language to imagine how unity and distinction can coexist without contradiction, and how God can bring forth multiple realities from His single, unbroken life.

🔬 📖 SPIRIT AND SCIENCE AS CO-LABORERS IN HUMAN DESTINY

At the foundation of existence, where biblical revelation and scientific discovery converge, lies a profound truth: the unseen continuously shapes, sustains, and redeems the visible. Whether examined through scripture or science, the cosmos reveals an active matrix of unseen forces from which all order arises. Far from standing in opposition, faith, and science function as twin lenses, illuminating different aspects of the same reality.

THE UNSEEN FORCES AFFECT CREATION

Enigmatic Threads

The blueprint of life is not just molecular—it is spiritual. Breath and design, quantum and covenant, all point toward a deeper merger. Let us now consider humanity as the sacred intersection of matter and meaning.

📖 Chapter Nine: The Fabric of the Unseen

📜 🪡 The Fabric of the Unseen: Woven by Forces Both Physical and Divine

The unseen does not mean the unreal, it is the foundation upon which everything stands. As we read before, the universe itself is a tapestry woven of invisible forces, gravity, electromagnetism, nuclear interactions, all humming beneath the surface, sculpting the cosmos. Yet, scripture tells of another unseen force, a presence beyond particles and equations. The Spirit of God, moving across the waters, breathing life into dust, guiding creation with precision unseen yet undeniable.

If the universe is a great orchestra, then physics describes the instruments and vibrations, while faith reveals the Composer. Could it be that both speak of the same symphony, one heard in waves and frequencies, and the other felt in spirit and purpose?

The Spirit does not dwell only in temples—it moves through every moment. In creativity, in healing, in relationship, the invisible becomes visible. Let us now explore how divine agency shapes the everyday.

`📖 Chapter Ten: The Omnipresence of Energy, the Omnipresence of God

Science tells us that electromagnetic waves flow endlessly, penetrating space, passing through matter, unseen yet ever-present. These waves carry voices, knowledge, power, illuminating darkness without ever being seen. And yet, scripture describes God's presence in the same way, invisible yet pervasive, flowing through creation, shaping reality, filling every corner of existence.

Like the electromagnetic spectrum, which exists even where the eye perceives only emptiness, so too does the Spirit, silent yet unceasing. The idea that creation is held together by forces beyond our senses is not a rejection of science, it is its deepest affirmation.

✨ 1. The Invisible Light Behind All Seeing

We often fail to notice the parallels between the physical world and the spiritual one, not because they are hidden, but because they are so constant that we stop seeing them. Science teaches that electromagnetic waves are omnipresent — always here, always moving through us, always invisible, yet making visibility possible and lighting up the darkness. In the same way, the invisible Spirit brings forth true light. He illuminates the darkness within us, the darkness of nonexistence before creation, and the darkness that lingers within existence itself. What science describes in the physical realm; the Spirit fulfills in the spiritual: the unseen presence that makes all seeing possible.

📖 Chapter Eleven: The Blueprint and the Breath

🧬 🪡 🧬

The blueprint of life is not just molecular—it is spiritual. Breath and design, quantum and covenant, all point toward a deeper merger. Let us now consider humanity as the sacred intersection of matter and meaning.

🧬 🪡 🧬

🧬 🪡 Invisible Forces: The Divine Blueprint and the Cosmic Thread

As mentioned previously, before the universe unfolded, the fundamental forces, gravity, electromagnetism, strong and weak nuclear forces, were compressed together in an unbroken unity. Then, in the first moments of existence, these forces separated, each taking its role in shaping time, matter, and motion.

In parallel, scripture describes creation as an intentional act, a design brought forth through divine breath. The Spirit hovered over the waters, speaking existence into being. Is it possible that just as physics tells of a great unfolding of forces, faith speaks of an unfolding of divine will, an act of separation, movement, expansion, forming the intricate balance of life?

Both narratives seek to understand the emergence of order from mystery, where the visible world was once hidden within the unseen.

Faith and Physics: Searching for the Same Light

Many embrace the unseen forces of physics, the quantum interactions flickering beneath reality, the hidden fields governing matter. Yet some hesitate to believe in the unseen presence of God, though it moves just as invisibly through existence.

Could it be that science and scripture are staring at the same truth, one describing the mechanics, and the other the meaning? Could the electromagnetic waves, the gravitational pull, the cosmic web be more than equations? Could they also be the fingerprints of divinity, stretching across the stars and into the human heart?

Perhaps the difference is not in whether the unseen exists, but how we choose to interpret it.

HUMANITY AS THE LOCUS OF THE MERGER

At the core of this divine-scientific unity is humanity, the unique link between spirit and matter. Genesis tells us, and biology agrees, that humans are made from the dust of the earth. But mere atoms don't create life; something more is needed. "God breathed into his nostrils the breath of life, and the man became a living soul." This isn't just a myth; it's the sacred union that defines our existence: matter filled with spirit, body given a soul.

Science also recognizes this mystery. Neuroscientists study electrical activity, neurotransmitters, and synapses, but none can pinpoint the source of consciousness, the essence of "I." The mystery of will, thought, and moral law can't be explained by biology alone. Instead, quantum physics offers an amazing parallel: observation itself turns possibilities into reality. Here, Christ's words resonate deeply: "If you have faith as small as a mustard seed…nothing will be impossible for you" (Matthew 17:20). Just as observation solidifies reality from endless possibilities, faith, no matter how small, anchors unseen potential into existence. Could this reflect the unseen power of human decision and conscience, the realm of will and spirit?

Evolutionary biology shows the development of life, but it can't explain the universal longing for meaning, redemption, and purpose. Here, science and scripture don't conflict but complement each other, revealing humanity as the meeting point of physical existence and unseen intention. Faith, like observation, acts as the force that turns possibility into tangible reality, illuminating the divine mechanism that shapes both the physical and spiritual world.

📖 Chapter Twelve: The Spirit in The Everyday

The sacred is not confined to sanctuaries. It moves through kitchens, classrooms, and quiet moments. Let us now explore how the Spirit breathes through the ordinary, revealing the divine in daily life.

🌿 💡 THE SPIRIT IN THE EVERYDAY, INVISIBLE AGENCY IN HUMAN LIFE

The merger of the visible and invisible is not a distant idea; it plays out in everyday life.

Creativity: What is the sudden flash of insight, the poet's phrase, the mathematician's eureka moment, but the crossing of boundaries, where the unseen manifests into form? Theology calls it inspiration, psychology labels its intuition, but neither deny that true innovation comes from beyond accessible mechanisms.

Society & Relationships: Sociology describes the invisible bonds of trust, love, and shared values that hold civilizations together, far more powerful than contracts or economic exchange. The Bible affirms, "How good and pleasant it is when brethren dwell together in unity!" These unseen connections shape justice, sacrifice, and loyalty as real forces that govern human interactions.

Healing & Medicine: Science now confirms that forgiveness, gratitude, and hope impact therapeutic outcomes. Physicians observe the measurable effects of prayer, faith, and resilience on recovery. Love, faith, and hope, once considered immaterial, are now recognized as forces that shape biological well-being. The biblical call to "love one another deeply, from the heart" is echoed in modern medicine's realization that emotional and spiritual healing are as real as medical intervention.

Forgiveness, Memory, and Repeated Hurt

Forgiveness can be sincere long before the heart feels healed. A person may truly want to release the hurt, pray about it, and choose forgiveness with all their will, yet the memory of the wound continues to rise uninvited. This happens because the mind stores deep pain in a place meant to protect us, not punish the other person. When the memory resurfaces, it often spills out in repeated conversations, not to reopen the wound but to make sense of what still feels unresolved. The intention is not to harm, yet the repetition can bring anguish to the one who caused the hurt. This tension reveals a simple truth: forgiveness is an act of the spirit, but healing is a process of the heart. Until the wound is fully tended, the memory will echo. What breaks the cycle is not silence, but deeper healing — the slow rebuilding of trust, safety, and connection that allows the memory to lose its sting.

A New Way of Seeing the World

Stepping through the door does more than settle our destiny — it changes how we see. We remain in the world physically, but we are no longer of the world in identity. Yet in another sense, we become profoundly of the world in mission. Salvation moves us from the world's visibility into God's invisibility — from seeing only what is seen to seeing through the unseen reality of the Spirit. And once our sight changes, we are sent back into the world with a new purpose. We do not belong to the world's way of seeing, but we engage the world with God's way of seeing. We are not shaped by the world, but we are sent into it to shape it with the Gospel.

Choice is not neutral—it ripples through reality. From Eden to ethics, from prayer to sacrifice, the unseen consequences of human will shape the world. Let us now trace the threads of decision and destiny.

📖 Chapter Thirteen: The Threads of Human Choice

🕰🧠The Threads of Human Choice, Ripples Through Reality

Every choice we make ripples beyond the visible world. The idea of free will, first depicted in Eden, extends across psychology, ethics, and law. Science confirms that even small actions lead to wide-reaching effects, from genetic adaptation to environmental impact.

But unseen choices carry even deeper consequences. Silent prayers, acts of repentance, personal sacrifices, these shape human lives in ways that science cannot quantify. The Bible declares, "The effective, fervent prayer of a righteous person avails much." Shouldn't we recognize, and participate in, this invisible force shaping the world around us?

One Door, One Reality, Many Choices

To understand the power of this doorway, we must first recognize the two eternal possibilities that exist before anyone steps through it. Outside the door of God's salvation, two possibilities always remain: to stay outside or to enter in. But the moment a person steps through the door opened by Christ, the entire landscape of possibility changes. Just as a quantum system collapses into a single stable state once observed, entering into Christ collapses the probability curve of eternal outcomes. Jesus promises:

> *"My sheep listen to my voice; I know them, and they follow me. I give them eternal life, and they shall never perish; no one will snatch them out of my hand. John 10: 27-28*

The eternal question is settled. Yet within this secure reality, our daily choices still shape the life we experience. We no longer choose where we belong, but we continue to choose how we walk within the grace-filled reality we have entered.

A New Way of Seeing the World

Stepping through the door does more than settle our destiny — it changes how we see. We remain in the world physically, but we are no longer of the world in identity. Yet in another sense, we become profoundly of the world in mission. Salvation moves us from the world's visibility into God's invisibility — from seeing only what is seen to seeing through the unseen reality of the Spirit. And once our sight changes, we are sent back into the world with a new purpose. We do not belong to the world's way of seeing, but we engage the world with God's way of seeing. We are not shaped by the world, but we are sent into it to shape it with the Gospel.

Every choice sends a thread into eternity.

📖 Chapter Fourteen: Language, Covenant, and the Shared Invisible

🗣️🎶🗣️

Language is more than sound, it is sacrament. Through words, symbols, and shared rituals, the invisible becomes tangible. Let us now consider how covenant and liturgy bind us to one another and to God.

🗣️🎶🗣️

🗣️🎶 LANGUAGE, SYMBOL, AND LITURGY, MAKING THE INVISIBLE VISIBLE

Language and rituals help us understand hidden realities:

Speech: Invisible thoughts become audible sounds, shaping our perceptions and history.

Sacrament: Communion turns simple elements into

channels of grace, unity, and renewal.

Art & Music: The resonance of soundwaves stirs the soul, lifting and transforming the heart in ways science alone cannot explain.

Scripture teaches, "The kingdom of God is not a matter of eating and drinking, but of righteousness, peace, and joy in the Holy Spirit." Without the unseen, the visible becomes empty. Without the visible, the unseen remains mute and unshared.

⚖️ 💧 Justice, Compassion, and Societal Evolution, The Spirit Moves Through History

History itself is a tapestry of seen and unseen forces. Moses stood before Pharaoh armed only with God's invisible command, and Egypt was reshaped. Prophets spoke unseen hope that materialized generations later. The justice movement, from abolition to civil rights, began not with weapons, but with visions of a better world, driven by an unseen force of hope and conviction.

Science, when aligned with the Spirit, becomes a tool for redemption, curing diseases, lifting millions from poverty, expanding human understanding. Yet motivation, the sacrificial drive for justice, is never reducible to neurons or evolutionary survival. It is the Spirit's mark, the fingerprint of God pressing humanity beyond its limits.

Marriage, Family, and Covenant, The Visible and Invisible Sanctuary

Even the most personal spheres, love, marriage, and family, carry an unseen depth beyond biology. Science can explain genetics and attraction, but it cannot account for lifelong covenant, where forgiveness bridges failures, and fidelity transcends fleeting passion.

Marriage, family, and community endure not simply because of material bonds, but because unseen commitments, faith, sacrifice, and devotion, hold them together. The biblical covenant mirrors this in every enduring relationship: the visible bond depends on invisible vows, sustained by the unseen Spirit.

"To deny the invisible is to dismiss the foundation of reality. To embrace it is to step fully into wonder."

Choice is not neutral—it ripples through reality. From Eden to ethics, from prayer to sacrifice, the unseen consequences of human will shape the world. Let us now trace the threads of decision and destiny.

📖 Chapter Fifteen: Suffering, Resurrection, and the Call to Integration

Suffering is not the end—it is the threshold. Through pain, the Spirit speaks, and through resurrection, the invisible triumphs. Let us now enter the mystery of redemption and renewal.

SUFFERING, MEANING, AND ULTIMATE REDEMPTION

Suffering is where the unity of spirit and science is most deeply revealed. The Cross stands at the center of history, a visible tragedy, yet an invisible victory of sacrificial love. Psychology confirms that meaning, faith, and purpose transform trauma, turning suffering into resilience and testimony.

The Spirit does not erase pain, but fills it with purpose, transmuting loss into sanctification.

Resurrection and Transformation, The Climax of Unity

The final merger of the visible and invisible is found in the Resurrection, where matter was transformed, and Spirit became fully manifest. Science proclaims that matter cannot be destroyed, only changed. Faith declares that the same Spirit that animated the dust in Genesis will transform us in the end.

The mission of both science and faith is not reduction, but resurrection. Not merely explanation, but renewal. Not just logic, but life itself.

The Call to Integration, A New Era Dawning

Let it be clearly declared: to embrace either science or faith while rejecting the other is to diminish the fullness of truth. The biblical worldview has never been threatened by honest inquiry; on the contrary, every deepening of scientific knowledge only magnifies the awe and accountability of being co-creators, image-bearers, and stewards in a Spirit-soaked cosmos.

We stand at a threshold moment in history. Never before has science revealed so much invisible order, and never before has the human spirit been so restless for meaning. The task before us is not to choose between Genesis and genetics, between spirit and nature, but rather, to merge them.

To let the spiritual truths of purpose, dignity, and love breathe through the measurable details of nature. To see within each natural law a parable of the Living Lawgiver. To honor both reason and revelation, recognizing that the unseen holds together all that is seen.

This is the moment to step beyond division and into a unified vision of reality, where faith and science no longer compete but complement.

All threads now converge. The Spirit that hovered, the blueprint that breathes, the suffering that transforms — they form a weave that cannot be broken. Let us enter the final chapter, where the invisible becomes testimony, and the visible bears witness.

THE INFINITY WEAVE

Two interlaced threads form a continuous loop — the weave of Spirit and Science, eternity and breath. A symbol of the unbreakable pattern that holds creation together, inviting us to trace its threads.

📖 Chapter Sixteen: The Unbreakable Weave

The greatest evidence of truth is not theory—it is testimony. Changed lives, renewed hearts, and reconciled communities reveal the unbreakable weave. Let us now bear witness to the Spirit's work in flesh and bone.

🧠 🩶 🧠 The Most Convincing Testimony, Changed Lives

The greatest evidence of this divine-scientific unity is not found in textbooks or laboratories alone, but in human lives, renewed, redeemed, and transformed. The addict set free, the wounded forgiven, the community rebuilt, these are the visible fruits of invisible grace. Neuroscience can observe the rewiring of thought patterns, but faith proclaims the rebirth of the heart.

The merger is not theoretical nor abstract; it is flesh and bone, mind and spirit, world and Kingdom, reconciled and forever changed.

PERSONAL EXPERIENCE OF GOD' SPIRIT

Do I believe that God's invisible Spirit exists and is active in the world? Absolutely. I've experienced it throughout my life, even when I didn't recognize it at the time. Now, as I approach eighty years old, I can look back and see how God's invisible Spirit has guided and protected me as I stumbled through life. He performed miracles I didn't see at the time, opened doors that were closed, and created doors where there were none. This happened because of the prayers I didn't know were being prayed for me by my father, mother, and others.

I've read the Bible as an adult and taught Church School classes, but I don't remember many of the scriptures I've taught and read. However, as I reflect on my journey, one guiding principle stands out: what my parents taught me early on, The Golden Rule of Life: *"Do to others as you would have them do to you."* It took years for me to realize this originated from the Bible. Jesus speaks these words in Matthew 7:12: "So in everything, do to others what you would have them do to you, for this sums up the Law and the Prophets," and in Luke 6:31: "Do to others as you would have them do to you." I was living by the principles my parents taught me, not knowing it was God's invisible Spirit working in me.

Conclusion, The Unbreakable Weave

Within the inexhaustible depths where the Spirit of God and the discoveries of science entwine, there emerges a single, resounding call: Do not fear the mysteries. Pursue, with passion, both the visible and the invisible. Seek out both what can be measured and what must be believed. For in this merger, humanity fulfills its intended vocation, not as an accidental byproduct of randomness, nor as a frightened subject of impersonal forces, but as the visible image of the Invisible, called to steward, heal, co-create, and reveal the glory of the One in whom all things, seen and unseen, hold together. Such is the union of science and faith. Such is the destiny, dignity, and duty of every person who dares to see with both the light of reason and the fire of Spirit.

We are not merely formed of elements — we are kept alive by an invisible electrical presence. Science affirms what Scripture implies: that Spirit and Nature together awaken awareness. The weave is not just the structure of the universe, but the breath that animates it. Awareness is not a late arrival — it is eternal existence, activated through the interplay of soil, spark, and spirit, to be united with the Alpha of the Eternal, and to live united with the Omega.

So, I ask: If we trust the invisible forces that carry our words across space, and the unseen energies that sustain our bodies for a time we cannot predict — why then do we resist believing in the Invisible Force That offers life beyond time, transformation beyond decay, and eternity beyond breath?

The weave cannot be broken — it holds all things together, seen and unseen.

🧵 Motif Appendix A: Sacred Threads

A Poetic Journey Through the Invisible and the Visible

These motifs appeared throughout the journey from the invisible to the visible.

Each one is a thread in the weave — a pause, a breath, a revelation.

They are not decorations, but signals.

Not summaries, but stirrings.

Each motif marks a moment when the Spirit whispered beneath the surface of the text,

drawing attention to what cannot be seen,

yet must be felt.

They rise like echoes from Eden,

like sparks from the breath of God,

like fingerprints left on the fabric of creation.

To follow them is to trace the Spirit's movement —

through Scripture, through science, through suffering, through joy.

They are the sacred threads that hold the story together.

Before creation took shape,

before light pierced the void,

there was Spirit.

The biblical story begins not with form,

but with presence —

an invisible, living force

hovering over the waters.

The story begins not with matter,

but with meaning.

Before atoms formed,

before stars ignited,

there was a Word —

a divine intention

waiting to become visible.

Creation did not end with light.

It continued with the shaping of sky and sea,

the division of waters,

the forming of boundaries,

and the quiet choosing of lives

through whom God would work.

The Spirit's movement through creation

did not stop with nature —

it extended into the shaping of nations,

the rise and fall of empires,

and the whisper of prophecy

guiding history toward redemption.

Forgiveness is not the end of judgment —

it is the beginning of restoration.

Light reveals what darkness hides,

not to shame,

but to heal.

Where Spirit and science meet,

breath becomes bond,

and the invisible finds its equation.

Forces unseen —

fields, waves, resonance —

echo the deeper reality

of the Spirit who holds all things together.

Invisible forces shape all life:

the divine blueprint,

the quantum thread,

the coherence that binds

cell to cell,

heart to mind,

spirit to body.

Suffering, meaning, and ultimate redemption —

the arc bends toward light.

What breaks us

can also remake us,

and every shadow

becomes a canvas

for revelation.

The most convincing testimony

is a changed life.

No argument speaks louder

than transformation —

the invisible Spirit

made visible

in human hearts.

The weave cannot be broken —

it holds all things together,

seen and unseen.

Every thread,

every force,

every life

is part of a single tapestry

woven by the Spirit.

🌿 APPENDIX B

🧵 Sacred Threads: Symbol Index and Interpretive Guide

The motifs woven throughout this book serve as quiet guides, marking moments where the invisible becomes visible. Each symbol highlights a recurring pattern — a thread of meaning that rises from the unseen and shapes the story of creation, consciousness, and redemption.

These motifs help the reader trace the hidden architecture of the book. They echo the way unseen forces shape both the physical world and the spiritual one. Each symbol invites the reader to pause, look again, and recognize the unity beneath the diversity of themes.

🧵 🪡 🧵 The Sacred Thread – The Unseen Line Holding All Things Together

Meaning:

The continuous, invisible line woven through Scripture, science, creation, consciousness, and human experience. It embodies the book's central theme: the invisible becoming visible.

Appears in:

- Motif Appendix: Sacred Threads
- Part Three, Chapter Nine: The Fabric of the Unseen
- Part Three, Chapter Sixteen: The Unbreakable Weave

🌌 🌿 🌌 The Triple-Swirl — The Spirit Before All Things

Meaning:

The eternal presence of the Spirit — before creation, within creation, and beyond creation.

Appears in:

- Part One, Chapter One
- Part Three, Chapter Six

🌱 🌌 🌱 The Seed-Burst — Meaning Emerging from the Invisible

Meaning:

The emergence of intention, purpose, and moral reality from the unseen.

Appears in:

- Part One, Chapter Two
- Part Three, Chapter Five

🐑 🕊️ 🐑 The Shepherd-Spirit Motif — Guidance Through the Unseen

Meaning:

The Spirit's quiet work in shaping, guiding, and empowering a life from within.

Appears in:

- Part One, Chapter Three
- Part Three, Chapter Sixteen

The Sovereignty Motif — God's Invisible Governance

Meaning:
The Spirit's unseen influence over nations, rulers, and history.
Appears in:

- Part One, Chapter Four
- Part Three, Chapter Fourteen

The Mercy-Restoration Motif — Light Revealing Darkness

Meaning:
The Spirit exposes darkness not to condemn, but to restore.
Appears in:

- Part One, Chapter Two
- Part Three, Chapter Twelve

The Science-Spirit Convergence Motif — Where Physics Meets Revelation

Meaning:
The meeting place of scientific discovery and spiritual truth.
Appears in:

- Part Two, Chapter Nine
- Part Three, Chapter Ten
- Part Three, Chapter Eight

🧬 🧬 The Blueprint Motif — The Divine Pattern in Life

Meaning:
The invisible architecture of biological and cosmic order.
Appears in:

- Part Two, Chapter Five
- Part Three, Chapter Eleven

🌑 🌈 🌑 The Suffering-Redemption Motif — Darkness as a Canvas for Light

Meaning:
The arc from suffering to meaning, from darkness to transformation.
Appears in:

- Part Three, Chapter Fifteen

🧠 ❤️ 🧠 The Testimony Motif — The Visible Evidence of the Invisible Spirit

Meaning:
The changed life as the clearest sign of the Spirit's presence.
Appears in:

- Part Three, Chapter Sixteen

🔗 ✨ 🔗 The Unbreakable Weave Motif — The Unity of All Things

Meaning:

The final synthesis — the unity of science and Spirit, body and soul, matter and meaning.

Appears in:

- Part Three, Chapter Sixteen
- Closing of the Motif.

Symbols are the footprints of the unseen,

left for those who look twice.

APPENDIX C

🗝️ SIMPLIFIED MOTIF KEY

A quick-reference guide for readers who want a one-page summary of the book's symbolic language.

The Sacred Thread: (🧵) The invisible line connecting Scripture, science, and the human story.

The Triple-Swirl: (🌌 🌿 🌌) The Spirit's eternal presence before, within, and beyond creation.

The Seed-Burst: (🌱 🌌 🌱) Meaning and intention emerging from the unseen.

The Shepherd-Spirit: (🐑 🕊️ 🐑) The Spirit's quiet guidance and formation.

The Sovereignty Motif: (🏳️ 🏛️ 🏳️) God's unseen governance over history and nations.

The Mercy-Restoration Motif: (🕊️ 🌌 🕊️) Light revealing darkness for the sake of healing.

The Science-Spirit Convergence: (⚛️ 🏳️ ⚛️) Where scientific discovery mirrors spiritual truth.

The Blueprint Motif: (🧬 🗝️ 🧬) The divine pattern woven into life and creation.

The Suffering-Redemption Motif: (🌑 🌈 🌑) Darkness,

becoming the ground for transformation.

The Testimony Motif: (🧠❤️🧠) Changed lives as evidence of the Spirit's work.

The Unbreakable Weave: (🪢🌟🪢) The unity of all things in God's design.

The invisible is not far.

It is simply deeper.

Spirit, Soil, Spark

🔗 We Who Witness the Weave

We are not the weave.

We are the part of it

that has been given the honor to witness.

From nature, the earth, and the spirit,

we emerge —

not as authors of the pattern,

but as those invited to trace its threads.

The breath is not borrowed.

It is embedded.

Both spirit and soil

are always working from within.

The weave cannot be broken.

It is the Divine masterpiece —

vast, intricate, eternal.

But we —

we are the ones given the opportunity

to experience it,

to try to understand it,

to marvel at its complexity.

This is our privilege:

not to control the pattern,

but to glimpse its glory.

And so the question remains:

Who decides the victory?

Victory is not declared by dominance.

It is discerned in integration.

Not in the silencing of one thread,

but in the harmony of all.

Victory is when the weave holds —

when tension becomes tapestry,

when breath and body, science and soul,

do not compete, but complete.

Victory is not a finish line.

It is a revelation:

that the Divine has always been

working from within.

The weave holds.

And we, we who witness,

are given breath to see it.

The Return

The visible world softens at the edges,

its brightness returning to the field beneath it.

Matter becomes motion,

motion becomes meaning,

and meaning returns to Spirit.

Nothing vanishes; everything transforms.

What we once saw with our eyes

now lives in a realm too deep for sight,

yet nearer than breath

Visible to Invisible

All that is, and all that was, will be.

Listen, I tell you a mystery: "We will not all sleep, but we will all be changed, in a flash, in the twinkling of an eye, at the last trumpet. For the trumpet will sound, the dead will be raised imperishable, and we will be changed." *1 Corinthians 15:51-52*

In the language of thermodynamics, the First Law declares: $\Delta U = Q - W$

Energy is neither created nor destroyed — only transformed. The internal energy of a system changes as heat enters and work is done. But beneath this equation lies a deeper truth: that the universe itself is a vessel of divine conservation. The Spirit does not waste energy. Every breath, every loss, every exertion is accounted for in the sacred balance. Even in suffering, something is being transfigured. The unseen work of God is not subtraction, but transformation.

Mama

Dec 18, 1902 – Mar 21, 1993

I Sit, I See

I see beyond that – that is before me.

I see from that – that was, that – that is.

I see from that – that is, that – that will be.

Experiences through the years have bestowed wisdom and given insight into life through the totality of her experiences. Witnessing scientific and technological changes have destroyed, changed, and developed new philosophical and theological thoughts. She sees. She sees much more than just the reality of that which is before her, she sees from that – that was, the past, and the totality of that – that is, the present. Like an intricate mathematical function, the summation of each integral of that – that was and that – that \is, provides sight into that – that will be, the future.

Other books by the Author

God in Mathematics: The Novel Unseen Love

Hidden Away

Baby Power Why Are We Here? From Where Did We Come? Answered

Father Forgive Them for They Know Not What They Do

Too Much Pain

Created for His Glory: Who and Why?

Growing Up Colored Negro and Black in the Gap

God's Plan of Salvation

www.ingramcontent.com/pod-product-compliance
Lightning Source LLC
LaVergne TN
LVHW090404210926
848480LV00002B/441